COUNTING AND FINANCIAL LITERACY BOOK

Pre-K
1st Grade
Age 3 -7

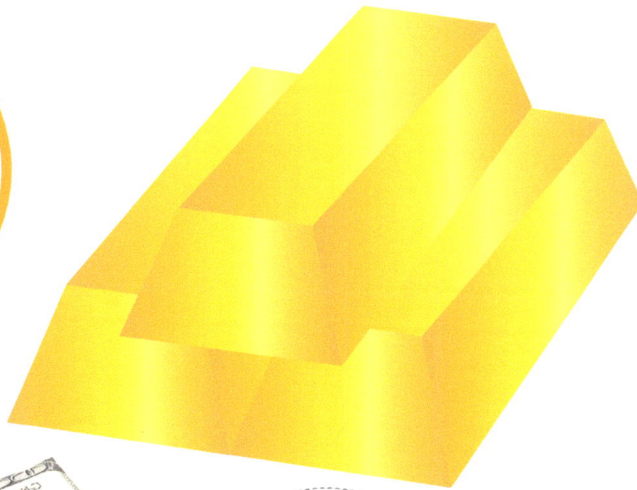

LITERATURE RAYS

First paperback edition

Book design by Kireeha S. Plair

ISBN: 978-1-7368783-7-8

Published by: Literature Rays

Email: Literaturerays80@aol.com

Mansa Musa
Born c.1280- Died c.1337

Enlightening facts

Mansa Musa was a king from mali. He is the riches man to ever live. He was said to have been very giving and earned most of his fortune through trading gold and salt in west Africa. He also sold ivory from elephants.

"Remember children to be creative and utilize all of your talents. Multiple streams of income are your building blocks to generational wealth."

-Kireeha plair

COUNTING BY ONES

(1 quarter)	
(2 pennies)	
(3 dimes)	
(4 cents)	
(5 five pence)	
(6 dollar coins)	
(7 nickels)	
(8 quarters)	
(9 dimes)	
(10 pennies)	

COUNTING BY 2'S

COUNTING BY 5'S

COUNTING BY 10'S

Counting American Coins

Count the amounts in the jars, and write the total underneath.

_____

Counting American Coins

Count the amounts in the jars, and write the total underneath.

_____

Counting American Coins

Count the amounts in the jars, and write the total underneath.

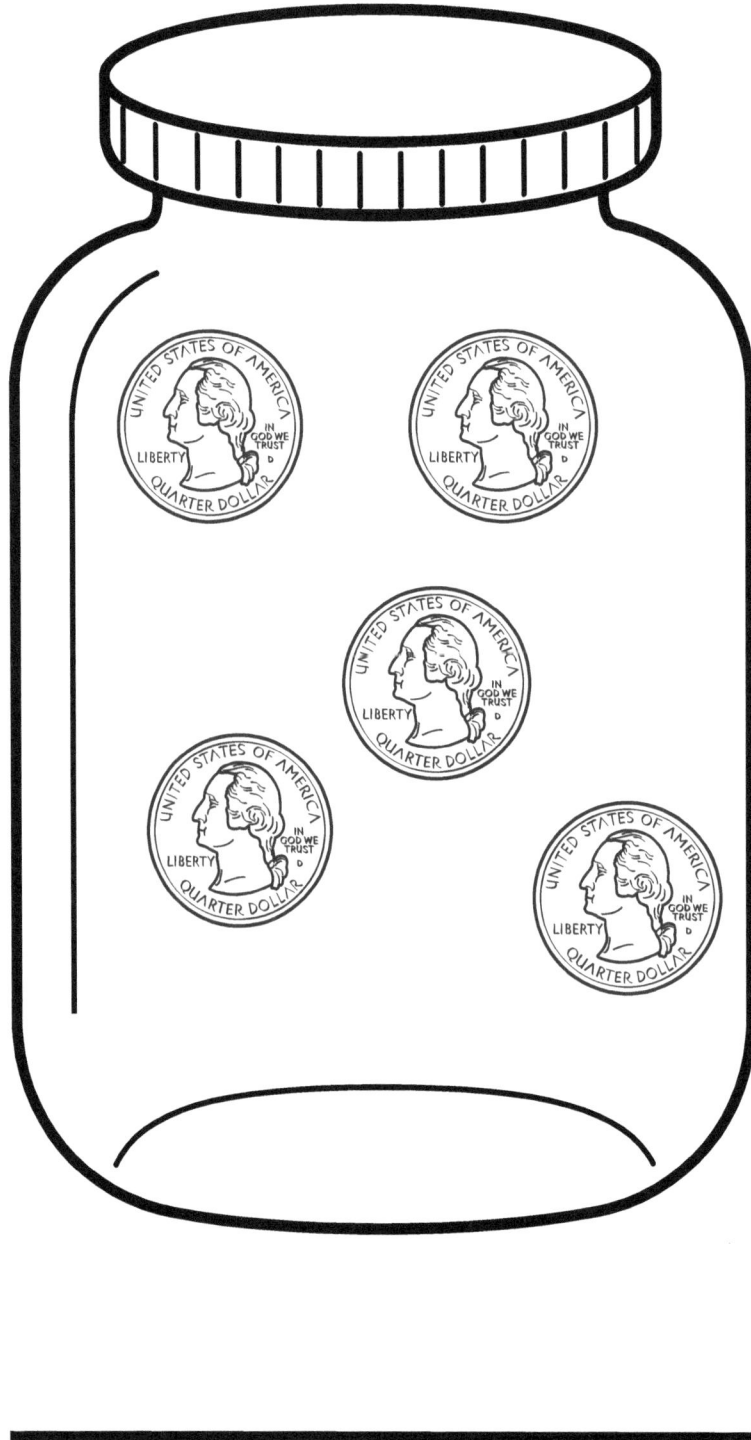

_____

Counting American Coins

Count the amounts in the jars, and write the total underneath.

_____

Counting American Coins

Count the amounts in the jars, and write the total underneath.

_____

Counting American Coins

Count the amounts in the jars, and write the total underneath.

_____   _____   _____   _____

_____   _____   _____   _____

COUNTING MONEY

Add the coins and write the total in the box.

  =

   =

   =

 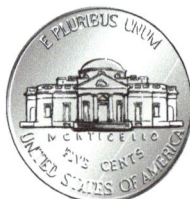  =

    =

    =

PENNIES AND NICKELS

Count the pennies in the left and write the amount in nickels.

_____ =

_____ =

_____ =

_____ =

Price tags

Cut out, place on items and play shops using play money.

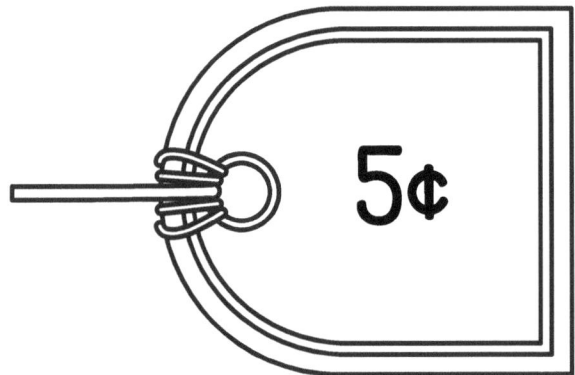

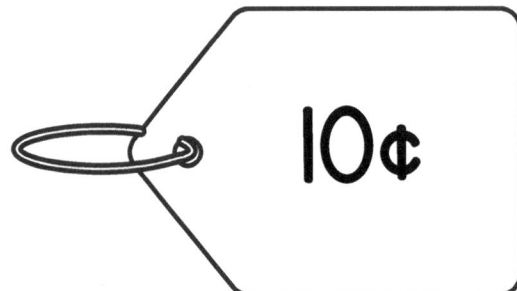

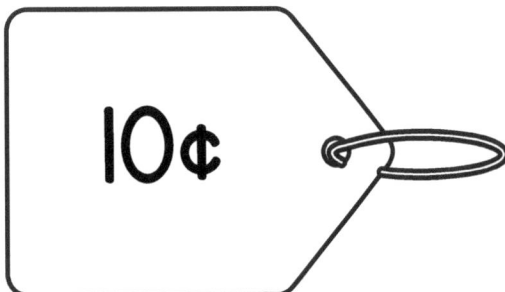

$1

20¢

$2

5¢

50¢

10¢

10¢

$2

MONEY ADDITION

Count the number of times each bill appears, then calculate the total for each bill:

	$1	$2	$5	$10	$20	$50	$100
Total Count							

 # Money Match

Match each amount of coins with the correct amount of money.

   • 30¢

   • 37¢

   • 5¢

   • 10¢

   • 17¢

   • 15¢

American Money- Coins

Write the correct total to each set of coins.

¢0.26   ¢0.08   ¢0.70   ¢0.46   ¢0.65

American bills and Coins

PENNY

1¢   1 cent $.01

NICKEL

5¢   5 cent $.05

DIME

10¢   10 cent $.10

QUARTER

25¢   25 cent $.25

HALF DOLLAR

50¢   50 cent $.50

DOLLAR BILL

1 dollar $1.00

5 DOLLAR BILL

5 dollar $5.00

10 DOLLAR BILL

10 dollar $10.00

20 DOLLAR BILL

20 dollar $20.00

50 DOLLAR BILL

50 dollar $50.00

100 DOLLAR BILL

100 dollar $100.00

1'S MULTIPLICATION

1 x 1 = 1

2 x 1 =

3 x 1 =

4 x 1 =

5 x 1 =

6 x 1 =

7 x 1 =

8 x 1 =

9 x 1 =

10 x 1 =

2'S MULTIPLICATION

1 x 2 = 2

2 x 2 =

3 x 2 =

4 x 2 =

5 x 2 =

6 x 2 =

7 x 2 =

8 x 2 =

9 x 2 =

10 x 2 =

5'S MULTIPLICATION

1 x 5 = 5

2 x 5 =

3 x 5 =

4 x 5 =

5 x 5 =

6 x 5 =

7 x 5 =

8 x 5 =

9 x 5 =

10 x 5 =

10'S MULTIPLICATION

1 x 10 = 10

2 x 10 =

3 x 10 =

4 x 10 =

5 x 10 =

6 x 10 =

7 x 10 =

8 x 10 =

9 x 10 =

10 x 10 =

Money Words

Write the money words.

quarter

1. --------------------------------------

 nickel

2. --------------------------------------

dime

3. --------------------------------------

 penny

4. --------------------------------------

dollar

5. --------------------------------------

 half dollar

6. --------------------------------------

QUARTER

There are four quarters in one dollar.

Quarter
25 cents

quarter

quarter

A quarter

is 25 cents.

DIME

There are ten dimes in one dollar.

Dime
10 cents

dime dime

dime dime

A dime

is 10 cents.

PENNY

There are one hundred pennies in one dollar.

Penny
1 cent

penny

penny

A penny

is 1 cent.

NICKEL

There are twenty nickels in one dollar.

Nickel
5 cents

nickel

nickel

A nickel

is 5 cents.

MONEY
MONEY
MONEY

Instructions:

- Draw a line to match the American coins and currency to their value.

10$

$1

$5

25¢

10¢

5¢

American Coins

Draw a line to match the coin to its value!

10¢   25¢   5¢   1¢

How many of each coin can you spot?

_____   _____   _____   _____

COLOR THE COINS

Color the coins below using the colors provided in the key.

RED

BLUE

YELLOW

GREEN

IS MONEY ENOUGH TO BUY?

✓ Check the box if it is enough money to buy
the item. If not put a X in the box.

Item	Price	Money	Box
Pencil	3¢		☐
Shirt	$5		☐
Ice cream	50¢		☐
Watch	$70		☐
Candy	2¢		☐

ICE CREAM STORE

Color the coin to pay for the Ice crem.

¢.25	¢.10	¢.05
¢.50	¢.25	¢.05
¢.10	¢.20	¢.10

DEFINITION

$ symbol means dollars

¢ symbol means cent (change/coins)

Savings	The portion of money earned that is not spent.
Good savings habits	Every dollar earned save a quarter from it. Start saving as soon as you make your first dollar. Never spend out of your savings unless you have no other choice.
Budget	A plan on how you will spend and save your money. Budget rules: 50/25/25 Spend 50 percent of your earnings on needs 25 percent on wants 25 percent on savings
Credit	Credit helps people to purchase goods or services using borrowed money, The lender expects to receive the payment back with extra money (called interest) after a certain amount of time.
Credit utilization	Your credit utilization ratio should be 30% or less, and the lower you can get it, the better it is for your credit score. Your credit utilization ratio is one of the most important factors of your credit score.
Invest	Take some of the money earned to pay into a source that will make your money work for you. Example stocks,real-estate, ideas ect.
Lender	Lender is a person or business that lets you borrow money or a service up front and allows you to pay them back later.